Mary C. Sloan Woodward

Roses and Thorns

Mary C. Sloan Woodward

Roses and Thorns

ISBN/EAN: 9783337419981

Printed in Europe, USA, Canada, Australia, Japan

Cover: Foto ©berggeist007 / pixelio.de

More available books at **www.hansebooks.com**

ROSES AND THORNS

BY

MARY C. SLOAN WOODWARD

ILLUSTRATED

DAYTON, OHIO
PRESS OF UNITED BRETHREN PUBLISHING HOUSE
1894

To All My Friends,

BOTH HERE AND BEYOND THE RIVER,

These Poems

ARE

AFFECTIONATELY DEDICATED

BY

M. C. S. W.

Preface.

It is without any feelings of egotism or of self-importance that I offer these simple lines to the public, but with the hope that they will be cherished by my friends, and also with the belief that they will find an echo in other hearts.

I make no claim to the divine gift of poesy. My poems are unembellished and unadorned, and were written under many difficulties. The inheritance of defective vision has been a lifelong sorrow, and has made my work very arduous, and but for the aid of a kind sister, who has acted as my amanuensis, it would never have been accomplished.

My affliction has also prevented much of the revision of these poems which would otherwise have been performed. I therefore, asking the indulgence of my friends, offer them to the public as they are.

Mary C. Sloan Woodward.

Contents.

Roses and Thorns

Roses and Thorns.

MY WISH.

OH! may no sentiment of mine
Upon these pages record find
    Which prompts a thought impure;
May every line herein contained,
For guilelessness and truth unfeigned,
    Unsullied stand and pure.

May nothing I have written here,
Though it may cause a smile or tear,
    Call forth one word unkind;
May some emotion here expressed
Bring comfort to some heart distressed,
    Some soul new courage find.

'T will give me joy beyond the tomb
To know that in my earthly home
    Some token I have left
To aid some pilgrim on his way,
Shed o'er life's path some cheering ray,
    Or soothe some heart bereft.

THE WARNING.

A TEMPERANCE STORY.

IN a quiet and quaint little village,
  Nestled down among hillocks and dales,
Where the sweet, merry notes of the wild-bird
  Resounded through woodlands and vales,
There lived a fair, beautiful maiden,
  With eyes of a soft azure hue,
With tresses all sunny and golden,
  And heart ever loyal and true.

This maid had a handsome young lover,
  Whom she worshiped as almost divine;
And Albert was good, true, and noble,
  But, oh! sad to tell, he loved wine.
The fond mother warned and entreated;
  Her eyes saw a desolate hearth,
The wreck of two lives by the rum fiend,
  The years that the heart will have dearth.

"Dear mother, your fears are all groundless,"
  Said Helen; "I have not a fear,
For dear Albert loves me too truly
  To cause me a pang or a tear."

Her trusting heart saw not the shadow,
  Her dreams were all visions of bliss;
She sensed but the sunshine and brightness,
  Nor heeded the yawning abyss.

One day in the glorious autumn,
  When hillside and valley were bright,
And the many-hued tints of October
  Lent pictures to gladden the sight,
Fair Helen was led to the altar;
  Her heart sang a gladsome refrain,
The wedding bells rang out the chorus,
  And echo repeated the strain.

They left the dear homes of their childhood,
  With all the delights youth had known,
And hied to a far Western city,
  To plant a roof-tree all their own.
Fond parents, with doubts and misgivings,
  Bowed to the stern mandates of fate;
The future lay shrouded in darkness—
  They could only be patient and wait.

Years passed; and the beautiful Helen,
  Once happy, light-hearted, and gay,
Bore traces of sorrow and anguish;
  The roses had faded away

From her cheeks, and her step, once so buoyant,
  Was weary, and languid, and slow;
For Albert was cold and neglectful,
  And cared not, nor heeded her woe.

The serpent had coiled close around him;
  His manhood, his reason,—his all
Was fast in the toils of the viper
  That held all his being in thrall;
The chalice once filled with life's nectar
  Was broken and dashed to the ground;
The sun had gone down at the noontide,
  And vanished in darkness profound.

One night, after long, tearful waiting,
  A burden was borne to her door—
The wreck of her once noble husband
  Laid bleeding and cold on the floor;
In that moment of bitterest anguish,
  Let us hope the dear angels above
Poured into that heart, crushed and stricken,
  The balm of their pitying love.

The work of the demon was finished,
  The grave had closed over her dead;
She gathered her little ones 'round her,
  And bathed with tears each shining head;
"We must go now," she said, "my poor darlings,
  From these scenes of sorrow depart;

We must all go and ask your dear grandma
  For a place in her home and her heart."

She disposed of her scanty belongings,
  Bade friends and kind neighbors good-by,
And embarked on the long, weary journey,
  To the dear home of childhood to fly;
The goal was at length reached in safety,
  And as fond parents came forth to greet
And welcome their long-absent daughter,
  She fainted and fell at their feet.

She revived, but fond love and affection,
  Though soothing the anguish and smart,
Could not give back the life, wrecked and broken,
  Or heal the poor, bruised, bleeding heart;
She faded away like a lily
  From the stem rudely broken and torn,
And her little ones, now doubly orphaned,
  Are left in this cold world to mourn.

Without a kind father's protection,
  No mother's fond love and caress—
God pity the dear orphaned children,
  Alone in earth's dark wilderness.
They made her a grave in the hillside,
  And daisies in springtime will come
To hallow the tomb where the lone one—
  Fair Helen—lies, murdered by rum.

Fair maidens! Oh, list to the warning
  That bids you be wise and beware!
Oh, pause at the terrible vortex,
  Ere fate drags you down to despair!
Oh, pause! In the name of the millions
  Whose blood cries aloud from the tomb,
Whose lives have been laid on the altar,
  And crushed out in sorrow and gloom;

In the name of the grief-stricken mothers
  Who weep o'er their lost ones to-night;
In the name of the wives whose lone vigils
  Fade into the gray morning light;
In the name of the innocent children,
  Half famished, and ragged, and cold,
Whose fathers are wasting their substance
  To add to the rumseller's gold,

Don't marry a man to reform him,
  If ever so loyal and true;
The love that goes erst to the wine-cup
  Will soon prove inconstant to you;
The Upas holds fast to its victims,
  Persuasion and reason are vain,
But madly they rush to the maelstrom,
  Where multiplied thousands are slain.

O women and men of Columbia!
  As proudly and grandly it waves,

"There lived a fair, beautiful maiden,
With eyes of a soft azure hue."

The boast of American freedom
  Floats over a nation of slaves;
By passion and appetite fettered,
  Oh, dare not to call yourselves free!
While bound hand and foot by the rum curse,
  True freemen you never can be.

Oh, cast off the shackles that bind you,
  And stand forth untrammeled and strong!
Oh, stamp out this infamous traffic,
  This body- and soul-crushing wrong!
Oh. rise in the strength of your manhood,
  And wash from our dear native land
The blot from her honored escutcheon,
  That stainless and free she may stand!

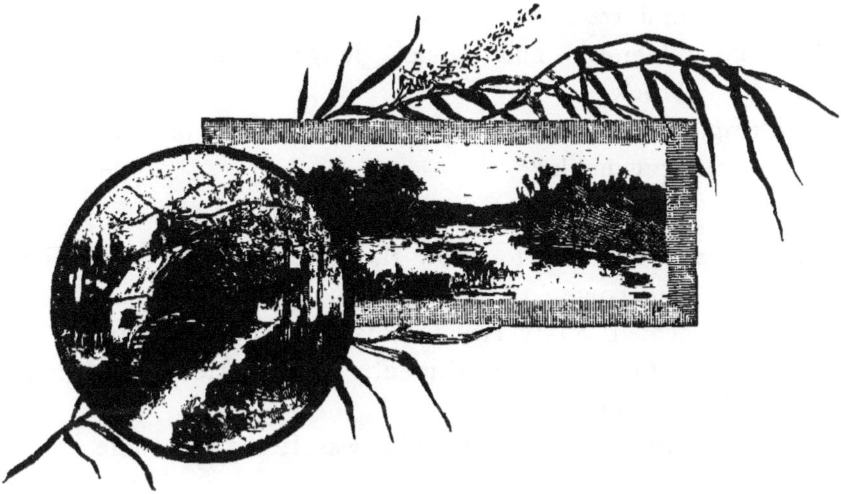

A SUMMER'S OUTING.

THE birds are singing sweetly, and the sky is fair and bright;
The sun smiles down upon the earth in streams of golden light;
But it somehow seems so lonely on our quiet street to-day,
For we've lost our pleasant neighbors who lived just across the way.

They have been here all the summer, drinking in the balmy breeze
Wafted from the fields and hillsides, from the fragrant flowers and trees;
Plucking golden-rod and daisies, 'long the highway, by the brook;
Gath'ring health, and strength, and freshness in each quiet, cozy nook.

And the way we came to know them was in this wise: They had come
From the heated, dusty city to this pleasant rural home,
Seeking comfort and refreshment, with a friend they long had known,
A young lady of our village, who was living all alone.

As to Newport, and Bar Harbor, Narragansett, and the rest
Of the fashionable places some enjoy with so much zest,
Where the strength is all exhausted in a ceaseless round of dress,
Each one vying with the other to excel in loveliness,—

There is naught of real enjoyment in a summer home like this;
And our friends were wise in choosing one of restful quietness,
'Neath the evergreens and maples, where the grasses fresh and sweet
Form a verdant velvet carpet for the merry childish feet.

With them came two little children—one an infant, frail and fair,
One a winsome, blue-eyed fairy, with a wealth of sunny hair;
For a time they all were happy, and the hours passed gayly by;
But the boatman pale was waiting, and his barque was anchored nigh.

The little one was fragile, and he faded like a flower;
His spirit passed beyond the veil in life's brief morning hour;
In his little snow-white casket he was gently laid to rest,
Ere earth's paths of sin and sorrow his infantine feet had pressed.

'T is to me a pleasing fancy—perhaps some may think it queer—
That the children grow in heaven much the same as they do here;
That the little ones we 've laid away, with bitter tears, to rest
May be grown to perfect stature when we clasp them to our breast.

The season is approaching when the crimson tints appear
On the woodbine and the maple; and the time is drawing near
When tourists seek their city homes, bid woods and fields good-by,
Cease from rural joys and pleasures, for the autumn days are nigh.

The birds still warble sweetly, and the sky is fair and bright,
And Sol comes forth each morn and bathes the earth in mellow light;
But it seems so very lonely on our quiet street to-day,
For we miss our pleasant neighbors who lived just across the way.

"The season is approaching when the crimson tints appear."

INSCRIBED TO THE BEREAVED.

DECEMBER, 1886.

EACH blossom that with fragrance fills the air
Blooms only to decay;
And all we love, however bright and fair,
Must quickly pass away.

Death thrusts his sickle forth, rich harvest reaping,
And at his stern behest
We yield our loved ones to his silent keeping,
In his cold arms to rest.

Remorseless grave! Were there no Life Immortal,
No Everlasting Shore,
How could we stand upon thy gloomy portal,
And know that nevermore

The voices of the lost, in tender greeting,
Shall fall upon the ear;
That all of earthly joy, so frail and fleeting,
We leave forever here!

Be still, sad heart, and calm thy grief and sorrow,
Thy anguish and unrest;
And from this thought sweet consolation borrow:
"Whatever is, is best."

And though the treasures that we fondly cherish
Lie 'neath the Christmas snows,
And o'er our hopes thus early doomed to perish
The voiceless tomb doth close,

They are not dead; they only waiting stand
To give us welcome sweet,
And to the glories of the Morning Land
To guide our weary feet.

Beyond these gloomy shades, a glorious vision,
Undimmed by sorrow's night,
Shall greet us on the plains of joy elysian,
Radiant with Love's own light.

For all life's weariness and tribulation,
Its bitterness and tears,
We shall have rich and ample compensation
Beyond the starry spheres.

TRUTH.

THERE is a jewel of celestial birth
I day by day
Am seeking—'t is a pearl of priceless worth.
O gem divine! come thou and dwell on earth,
And with me stay.

This priceless pearl adorns the locks of age
And brow of youth,
It dwells with the unlettered and the sage,
Lives in the heart of king and humble page,—
Men call it Truth.

It tarries with the true and earnest souls
Who dare defend
The right—though tide of persecution rolls,
Life's ship be stranded on the treacherous shoals,
Ere comes the end.

O Truth, thou sacred amulet divine,
Abide with me!
I bow in adoration at thy shrine;
Oh, let thy coronet my brow entwine,
My day-star be!

IN MEMORY OF GRANDMA S.,

WHO PASSED BEYOND, NOVEMBER 13, 1892, AT THE AGE OF 94 YEARS.

ANOTHER soul has passed from anguish mortal,
Another sweet release;
Another entrance to the heavenly portal,
Where dwelleth rest and peace.

Another hears at last the welcome greeting,
Receives the guerdon blest;
Another weary heart has ceased its beating,
And finds a peaceful rest.

The life so full of purity and sweetness,
Veiled now to mortal sight,
Shall enter into new and full completeness,
Beyond earth's storms and blight.

Long years she sat in helplessness and blindness,
Waiting the summons home,
Still trusting that the Father's loving kindness
Ere long would bid her come.

Enfranchised now, with naught to dim her vision,
With form erect and strong,

She walks with loved ones on the plains elysian,
And sings a joyful song.

Her presence, like a sweet aroma, lingers
Around her vacant chair,
As if sweet roses plucked by angel fingers
Had left their fragrance there.

TO TWIN GIRLS ON THEIR SIXTEENTH BIRTHDAY.

MAY old Father Time make the years bright and bonnie,
And bring many birthdays to Maudie and Maunie;
May your young lives be crowned with joy's golden-hued light;
May the years glide along full of hope and delight.

If the clouds intervening in darkness should frown,
May a rainbow, appear from your sky shining down,
Shedding luster and peace on your sweet girlhood's prime,
And to strong, noble womanhood's heights may you climb.

MY MAYFLOWER[1] PREMIUM.

I'M one of the Mayflower family, I've sailed with them many a day;
Till I find a ship more worthy, I always expect to stay;
For our barque has glided smoothly, without a gale or storm,
And the winds have been propitious, and the breezes soft and warm.

And to-day the postman brought me, as he paced his daily rounds,
Along with some Floral Beauties,[2] as pretty as could be found,
The cutest little box; 't was made of pasteboard stout and strong,
One and half inches thick and wide and just four inches long.

Five dainty packages of seed were first to meet my view,
Enough for a small flower garden, and vegetable, too;
What possibilities they hold! what latent sweetness sleeps
Within the small integument which safe from danger keeps.

Verbenas, phloxes, pansies, too, a royal feast will spread
To greet our eyes each morning as the garden paths we tread;

[1] *The Mayflower*, a floral journal.
[2] Floral catalogues.

And then tomatoes, Picture Rock, in summer we may eat;
And new wineberries later on will give a bounteous treat.

I look again — a tuberose bulb (Excelsior Pearl) is seen.
Methinks I scent the fragrant gems upon their axis green;
In fancy I in days to come their waxen buds behold,
And see, some balmy summer morn, their pearly cups unfold.

Three gladioli next I find. Those who have seen them know
The grand and gorgeous colorings which on their petals glow;
Afar above the trailing plants which deck the garden bed
She lifts erect, with queenly pride, and rears her stately head.

A little zephyranthes bulb comes last upon the scene.
It scarce had pressed the soft, moist earth before a leaf of green
Came pushing forth — I seem to see the yellow beauties blow;
Among the leaves of emerald green like golden gems they glow.

I keep all the little boxes, for I often love to send
A rare new plant or bulblet to some dear and cherished friend,
And Uncle Sam so kindly lends a house both safe and warm
To shelter all their journey through, secure from wind and storm.

THE MINISTER'S FAREWELL.

THE people to-day, at the sound of the bell,
Assembled to listen to words of farewell;
For the minister preached his last sermon to-day,
And to new, untried fields will soon hasten away.

He has lived for six years in our village, and served
His people acceptably,—never has swerved
From what he deemed duty,—was faithful and true
To his flock and the work he had chosen to do.

He was courteous in action and kindly in thought,
A gentleman always,—*all* preachers are not;
He moved round among us with dignified grace,
For all had kind greetings, and soon won a place

In the hearts of the people. He grew and advanced
In thought and in mind, until now it has chanced
That a new field, and wider and broader, is sought,
Which new strength will demand in the regions of thought.

I am not a church member, but have great respect
For all earnest believers, no matter what sect—
Or of no sect at all, if but Truth be the quest,
And all are content in her teachings to rest;

And what in this minister most I admired,
And for which my highest regard was inspired,
Was the kindness and courtesy always received
By those who a different doctrine believed.

And, if honest and earnest, why can't we extend
To all the same privilege,—deem each a friend?
But because we all differ in make-up and mold,
If we can't think alike, we are left in the cold.

Each blade of grass differs, each flower and each leaf;
'T is the order of nature,—and 't is my belief
That it ne'er was intended humanity should
Be to this an exception: God's laws are all good.

The rose and the lily together may bloom,
Delighting the sense with delicious perfume;
Like the grasses and grains, which for use all combine,
Yet of different fashion in Nature's design.

Who inspires high regard? Is it he not ashamed
Of his honest convictions, though censured and blamed,
Or the pious pretender, who seeks but to move
In the popular current, society's groove?

In the day when, unveiled, every spirit shall stand
At the bar of his conscience, God will not demand
What you may *believe*—what profession you make;
But what you have *done* for humanity's sake.

THE FAMILY POCKETBOOK; OR, GRANDMA GREY'S STORY.

"Come, Grandma, tell us a story," cried a party of maidens gay,
Who had gathered to spend the evening at the home of Grandma Grey;
"We know you can tell us something, in either prose or rhyme,
Which will help us enjoy the evening, about 'ye olden time.'"

"Yes, girls, I can tell a story, a tale of long ago;
It portrays a life of sorrow, but much that you ought to know,
And if I can do you good, dears,—for the history is my own,—
I can tell you much that I shrink from — that a mother should blush to own.

"Like you, I was once light-hearted, and happy, and young, and fair,
And my mirror often told me I had beautiful golden hair;
But the withered locks that you see, dears, and the furrowed cheek to-day
Record much of grief and sadness in the years that have fled away.

"I was not born to idle habits, but my parents were well
to do,
And had plenty for solid comfort, and some of the luxuries,
too;
My education was ample in the line of useful things,
And I spent three years in college, which always much culture brings.

"I took lessons in elocution, and painting and drawing learned,
And other refined employments to which my attention turned.
My life was full of enjoyment, with friendship and sunshine
blest,
But I fell in love with a farmer, and — well, you know the rest.

"So we rented a place in the country with brambles and weeds
o'ergrown,
But we thought if we managed rightly, sometime it might be
our own;
And you'll never know, girls, till you try it, how hard is the
farmer's life,
And I'm sure you will never envy his weary and care-worn
wife;

"For from four o'clock in the morning till after the set of
sun
Is a round of toil and labor, which is never wholly done.
It sounds very pleasant, talking of the pleasures of rural life,
But you'll find things very different if you're ever a farmer's
wife.

"But the work didn't hurt so badly—for then I was young
and strong—
As some other little matters that gradually came along
And shadowed my new existence and covered me with the shame
Of bitter humiliation, out of which much torment came.

"My wardrobe was well replenished when I left the dear
home nest,
For my mother saw what I needed and wisely managed the
rest;
And one and a thousand comforts which money alone will bring
Always came without the asking, and free as the flowers of
spring.

"Well, after I had been married a couple of years or more
I was sadly in need of money; for, in looking my outfit o'er,
I found I had darned and mended till patience had ceased
at last
To be reckoned among the virtues, and might be with the
vices classed.

"And when I asked John for money, he grumbled, and always
asked
What I did with the last he gave me, though a month or
more had passed;
And I argued the matter with him, but he thought he knew
it all,
And he would not yield to a woman,—no, not though the
heavens should fall.

"I had drudged in my husband's kitchen, I had managed with tact and skill,
And now I was nothing to him but a slave to obey his will.
I had done the washing and baking, had cooked, and churned, and now
It made me feel hard and bitter, and I took a solemn vow

"That I, a creature with reason, and heart, and mind, and soul,
Would not bow to my husband's mandates and yield him supreme control
Of my brain, and bone, and muscle, of my nerves, and strength, and will,
Because I was born a woman, with her mission to fulfill.

"My soul writhed under the torture, but I saw no way to break
The chains that were forged around me, and so I began to take
A look at the situation, and wonder what I could do,
By a strategetic movement, to carry a project through

"By which I could have some freedom, some produce to money turn
Without asking John, like a beggar, for what I had helped to earn.
So I kept my secret counsel, resolved what I could not do
In sunshine, I'd do in shadow, to carry my purpose through.

"I sold each week for the market the butter and eggs and
such—
The surplus of all these products, if ever so little or much,
And I was expected to furnish from out of my little store
The coffee and tea for the table and many things needed
more.

"And here I began my scheming, for this was my only chance
To try to outwit my husband; I saw at a single glance
I could buy cheap tea and coffee,—for we always had the
best,—
Could gather the best for market and give the home folks
the rest.

"John grumbled at my cooking, but for that I did not care—
I could afford to be a fox, if he would be a bear;
But he did not know the reason that the biscuits were flat and
sad,
Or why the coffee was muddy, or why the tea was bad
(For the biscuits were minus butter, and the coffee was minus
cream,
For these went to the market to further my little scheme),

"Or why we had breakfast bacon instead of delicious ham,
Or why he missed fried chicken, as tender as young spring
lamb;
For I sold all I could gather, and it netted me quite a sum,
Which made my young eyes glisten—but the word was
always mum.

"Thus for years I drudged and labored, growing bitter, and
stern, and cold,
For my dreams and my daily queries were, what could I
turn to gold?
And John had hoarded his money, and paid for the little farm,
And we might have both been happy, had our hearts been
true and warm;

"But my nature had grown morbid, as the years went gliding by,
And my sons were sour and sordid, my daughters crafty and
sly,
And you know my poor boy Charley, who in anger struck
his wife,
Was convicted of her murder, and was shut away for life.

"Oh, girls! don't enter blindly the matrimonial state;
There's naught in your whole existence, be the venture
small or great,
That's fraught with such weighty import, with pleasure or
pain so rife,
That holds such momentous issues, as choosing a mate for life·

"Talk all these matters over with your lovers, and don't
embark
Without a compass, or rudder, or helm to guide your bark;
Compare opinions freely, and earnestly seek to know
Whether your thoughts and feelings can in the same chan-
nel flow;

"For love, without sure foundation on which to build and rest,
Is an evanescent passion, which can make no existence blest;
And that which should be sustaining when with life's cares beset,
Ends only in disappointment, in bitterness and regret.

"And the rock on which I was broken, and which made my home a hell,
Gives many unhappy women the same sad tale to tell;
And till there is some division of the earnings of man and wife,
A married woman can never lead a happy, contented life.

"No man has a right to ask a maiden to be his wife
Without making her a partner in the family funds for life,
With a right to use her portion just as her reason chose,
And not drudge away existence for only her board and clothes.

"I had a dear friend in the city (she went home long ago),
And we sometimes talked of those matters, as sisters will, you know;
She had just the same experience, as women have to-day,
And this same old bugbear, *money*, took much of her peace away.

"So she reasoned the matter over, and her husband was kind and good,
And wanted to make her happy, as a thoughtful husband would,

And he gave to her the rental of a business house down
town,
Which brought her much of comfort and something to call
her own;
And she said, 'We are both more happy than under the old
régime,
And I think we hold each other in higher and more esteem.'

"Not till women are independent, and no longer submissive
slaves,
Can they make of their home the Eden for which their
nature craves;
And the garden where human flowers grow up and bud and
bloom
Should be free from the weeds of discord and yield only
sweet perfume.

"Not till homes are made pure and perfect will crimes and
vices cease;
When mothers exult in freedom, and abodes of love and
peace,
Will the race grow grand and noble, and honor and virtue
shine
On the brow of each man and maiden, like a coronet divine."

TRIBUTE TO THE MEMORY OF A DEPARTED BROTHER,

WHO PASSED BEYOND APRIL 10, 1863.

PURE, gentle spirit! Can it be that thou
Hast passed away from earth, and left us here
In this cold, selfish world, to struggle on
Alone? And have we gazed for the last time
Upon thy dear remains, and seen them hid
Forever from our sight in the dark tomb?
And is that gentle voice we 've heard so oft,
Falling in soft, sweet cadence on our ears,
Forever silent in the cold embrace
Of the grim monster, stern, relentless Death?
Can we believe it true? O pitying Heaven!
If thou hast balm in store for stricken ones,
Vouchsafe unto our crushed and bleeding hearts
Some consolation in this dark, dark hour
Of bitter sorrow.

Dear departed one!
Just as the swelling buds came bursting forth,
And flow'rets peeped from out their wintry beds,
Ere vernal bloom had gladdened Nature's face,
Thy spirit passed to an eternal spring

Of fadeless beauty; in thy manhood's pride
Cut rudely down and numbered with the dead,
Ere yet the peerless beauties of thy mind,
So richly stored with priceless gems of thought,
Were half unfolded.

Brother, thou hast gone
To meet the sure reward of virtuous life,
A life on which no darkened stain of vice
E'er cast its blighting shadow—spotless, pure,
Unsullied as the rainbow's glowing tints.

Oh, how thy memory, lost one, lingers yet
Around our hearts, forever there enshrined!
The springing grass, the woods, the birds, the flowers,
The passing breeze, the gently rippling stream,
All seem in language eloquent and sad
To speak of thee and whisper, "He has flown
To that bright realm where flowers immortal bloom,
And birds of paradise forever wave
Their golden pinions in the breeze of heaven;
Where spirits roam through all the wide expanse
Of vast creation, glide from star to star,
Enlarging every power of mind and soul,
And, lost in rapturous admiration, gaze
Upon the glories of the Infinite."

Then, while we wipe the ever-falling tear,
And plant the roses thou didst love so well

Upon thy lonely tomb, that hallowed spot,—
Hallowed by dearest, fondest memories
Of scenes forever fled,—we'll think of thee,
And breathe to Heaven a silent, fervent prayer
That in that world where happy spirits dwell
We all may meet, and mingle evermore
With those we've loved on earth. Till that blest hour,
Dear angel brother, truest friend, Farewell!

TO ————

THE things that you cannot see, dear,
  In this tangled maze of life,
That are hid away in the darkness
  In this world of turmoil and strife,
You shall see with open vision,
  When the thin veil is rent away,
And from out the gloomy shadows
  Will shine forth the perfect day.

You shall see why the cup of joy, dear,
  Was dashed from your lips away,
And the bitter dregs of sorrow
  You are fated to drink to-day;
Why you walk alone in the shadow,
  The sunlight of love withdrawn,
And the bitter mem'ries haunt you
  Of the days that are passed and gone;

Why they cover you like a pall, dear,
  This sad, lonely New-Year's Day,—
You shall see with undimmed vision
  "When the mists have cleared away."
You shall know in the home of the angels
  Why all this torturing grief

Has been crowded into your being,
  And there find a sweet relief.

You shall there meet a sweet affection,
  Undying, and true, and pure,
Which will live through the future ages
  And steadfast for aye endure—
A love that will fill your spirit
  With holy and restful peace,
Which will drive out distrust forever,
  And bid every discord cease.

The purest and noblest souls, dear,
  Through the smoke and flame have passed;
Have been purged by the fires of sorrow,
  And have stood undefiled at last.
Await, then, with patient courage,
  Be life's journey short or long,
And may angel guardians lead you,
  And say to your soul, "Be strong."

THE AQUILEGIA (COLUMBINE).

AMONG all fair Flora's treasures
  Which such loveliness combine,
For our Nation's floral banner
  I would paint the columbine;
For around about my dwelling
  All the dry, hot summer through
Bloom these gems of grace and beauty—
  Bells of gold, and white, and blue.

There is beautiful Chrysantha,
  With its dainty cups of gold,
Which looks up each morn to greet me,
  Its new beauties to unfold.
How I love their dewy freshness,
  And their charming, airy grace,
As they gleam like golden sunbeams
  Streaming o'er fair Nature's face!

Then comes lovely, chaste Cerulea,
  Robed in heaven's glorious blue—
Color sacred long to friendship
  And affection pure and true;
Sacred to the brave defenders
  Of our Nation, tried and true;

Fell on many a field of battle
  Loyal hearts that wore the blue.

Alba, clothed in pearly whiteness,
  With a face so sweet and fair,
Gazes up to kiss the sunlight,
  Pure as robes that angels wear.
Flora's many beauteous treasures
  All much loveliness combine,
But upon our floral banner
  Paint the graceful columbine.

When my beauties sink to slumber
  Under winter's shroud of snow,
I will patiently await them;
  For in springtime's genial glow
I shall see their green leaves peeping,
  And again, with verdure sweet,
They will lift their peerless blossoms
  Our admiring eyes to greet.

I will plant and grow the darlings,
  And will scatter all around,
That among my floral sisters
  All their beauties may abound;
And when death at last and stillness
  Claim this form, they may entwine
For my brow some memory's chaplet
  Of the lovely columbine.

IN MEMORIAM J. L. M.

DEDICATED, AND RESPECTFULLY SUBMITTED, TO THE EXCELSIOR SOCIETY OF OSBORN, OHIO.

ONCE more the Beautiful City
  Its gates has thrown open wide
To receive a mortal pilgrim,
  Borne over life's foaming tide;
And a purer, sweeter spirit
  Never stood on the starry floor
Where the sound of angel footsteps
  Makes melody evermore.

Methinks as she passed the portal,
  Upon her enraptured sight
There burst forth a glorious vision,
  All radiant with love's pure light;
And the arms of an angel mother
  Pressed close to her loving breast
And welcomed her cherished darling
  To the land of peace and rest.

We shall miss thy gentle presence,
  Thy bright smile, which always beamed

Alike on both friend and stranger,
  And like a rich jewel seemed,
And like the spring's mellow sunlight,
  Whose life-giving rays impart
A touch of returning verdure
  To nature's frozen heart.

Rest safely at home, loved sister,
  Life's trials and conflicts o'er;
"Earth has one less patient sufferer,
  And heaven gains an angel more";
We will cherish in fond remembrance,
  While sadly we drop a tear,
Enshrined in the heart's deep chambers,
  Thy memory sweet and dear.

WHAT OF TO-DAY?

WHAT seed have you sown to-day, dear,
  From which sweet flowers will spring;
Which to some lone heart may give hope and cheer,
  And freshness and sunshine bring;
Which may brighten the path of some troubled soul,
  As she gathers each fragrant flower;
Which may give relief to the burdened heart,
  And gladden some lonely hour?

Have you spoken kind words to-day, dear,
  Which have fallen like sunbeams bright
On some sorrowing mortal's shadowy path,
  And have made the darkness light;
Have aided to carry life's load of care,
  Brought peace to some heart distressed;
Have lighted with smiles the pallid cheek,
  And given to the weary rest?

Is some generous act to-day, dear,
  Inscribed on life's page for you,
Which may aid some soul on the downward way
  To rise to the pure and true?
By kindness oft you may beckon back
  A brother from sin's dark way;

Inspire with courage to turn aside,
  And strive for a brighter day.

Let every shining day, dear,
  Some kindly action see,
Some tones of love, which will thrill the heart
  Like angel minstrelsy;
And reckon as naught the day which finds
  No seeds of kindness sown,
No words which fall like the healing balm
  Before the light has flown.

PLANTING CUTTINGS OF GERANIUMS ON THANKSGIVING DAY.

A True Story.

One Thanksgiving morning, which dawned clear and bright,
And into each window streamed soft mellow light,
We thought that we would not have company come
But just have a nice, quiet dinner at home.

So when turkey was ready I put it to bake
And proceeded some other arrangements to make,
When I spied in the window a large "Evening Star"
Which had dropped its last blossom its beauty to mar.

I had potted in autumn and brought it within,
All full of nice buds, for I thought it a sin
To leave it a prey to Jack Frost's cruel hand
When a warm sunny window all buds would expand.

And now it was ready down cellar to go,
To take a long rest while the frost and the snow
Made fruitful the soil, that the spring, warm and bright,
Might open new beauties to gladden the sight.

But first, I intended to clip off its head
And take some nice cuttings for next summer's bed;
And as the last blossom had faded away,
I thought, "Why not do it on Thanksgiving Day?"

So I brought in a pot, washed it all clean and sweet,—
It measured six inches,—and down at its feet
I filled in with charcoal two inches or three
And put in pure sand, pressing down evenly.

Then I decapitated my plant large and tall;
It gave nine thrifty cuttings; I planted them all,
Pressing all closely in 'round the edge of the pot,
Then watered and gave them a warm, sheltered spot.

Many persons think cuttings root only in June,
In September, or August; but I very soon
In a warm kitchen window can coax them to grow
And get ready for spring while the winter winds blow.

And they didn't damp off as sometimes cuttings do,
But went right along bravely and every one grew;
And I saw in the future nine "Evening Stars" shine
Studded o'er with pure blossoms with green to combine

For a charming bouquet to send out to a friend,
Or in chambers of sickness and suffering to lend
A picture of brightness to charm away pain
And back to blest health woo the sick one again.

How dearly I love them, the beautiful flowers!
How many bright moments and calm happy hours
I have spent in their culture in days that are past,
Which shall live in my heart while my being shall last.

"How dearly I love them, the beautiful flowers!"

They smile up to greet me at morn and at eve,
And speak words of comfort and peace when I grieve;

And whatever else claims my portion of care,
My sweet floral darlings come in for a share.

Perhaps you think turkey was almost forgot.
Be this as it may, be assured he was not;
For he came from the oven all tender and sweet,
And fit for a queen or her consort to eat.

So we all enjoyed turkey, and cranberries, too;
And now will it not be surprising to you,
If I tell you no plants gave me finer display
Than the cuttings I planted on Thanksgiving Day?

IN MEMORY OF A BELOVED NIECE,

MABEL ALDA POWELL.

FAREWELL, sweet one! thy gentle spirit 's flown,
  Pure and unsullied, to a holier sphere;
Has launched into the fathomless unknown,
  And we must wait a little longer here.

Cut down in childhood's loveliness and pride,
  The dearest hopes, the brightest prospects riven,
O'erwhelmed at once by death's dark, rolling tide,
  And onward by its rushing torrent driven.

'T is Nature's law, and man can only bow
  In calm submission to the stern decree;
What seems so dark and so mysterious now,
  In heaven's refulgent light we soon shall see.

Peace, bleeding hearts! God doeth all things well;
  The precious child who for a time was given
Has left this darksome vale, where sorrows dwell,
  To mingle in the pure delights of heaven.

No sickness there, no graves, no funeral knell,
  No deep, low note of wailing for the dead,

No mother's tears, no father's sighs, which tell
  The depth of anguish for the joy that 's fled.

How blest to know that on this night of sorrow
  The glorious morning light of heaven shall shine;
To know that there shall come a glad to-morrow,
  Illumined with a radiance all divine.

Father, we leave our treasure in thy hand;
  With strength and courage aid us to endure,
Till we shall join her in the spirit land,
  From sin and all its woes to dwell secure.

A few short years, and then we hope to meet,
  And greet her on those bright elysian plains,
To mingle evermore in union sweet,
  Where purest love and joy perennial reigns.

TO A SCHOOL FRIEND,

On Sending Her a New-Year's Card.

Dear friend, be reminded by this simple token,
Though youth's sunny morn and its joys are no more,
That friendship's bright chain, ever strong and unbroken,
Unites our hearts still, as in sweets days of yore.

The old year has flown; and its pleasures and sadness
Are pictures suspended on memory's wall;
The new year may bring its full measure of gladness,
Or bring us the coffin, the bier, and the pall.

The old year is dead, but its shadow still lingers
At many a hearthstone; full many a heart
Has felt the rude touch of its pitiless fingers,
Or fallen a prey to its merciless dart.

We gather life's roses, but yet we remember
That roses have thorns, and we feel the sharp pain,
That June is soon followed by chilly December,
The spring's fragrant breath by the cold winter rain.

Whatever Time brings us, of joy or of sorrow,
Whatever the future may bear to our door,

The present is ours; let us trust for the morrow,
And patiently wait for what Fate has in store.

Away from this sphere, with its strife and commotion,
Away from its conflicts, its toils, and unrest,
We look for a haven beyond life's rough ocean,
Where earth's weary children find comfort and rest.

A VOICE FROM BEYOND.

To ———.

YOU are sad to-day, my darling,
  For you 're thinking of the day
When I left the mortal casket
  And my spirit passed away.
While your bitter tears are falling,
  Tears for one you mourn as dead,
I am here to pour, like ointment,
  Benedictions on your head.

I would fain have walked beside you,
  And to shield you still have striven,
Sought to rear to noble manhood
  Those who to our home were given;
But the mandate which no mortal
  E'er had power to disobey
Rent our little band asunder,
  Filled your heart with sore dismay.

I am near to guard and cheer you,
  Though my form you cannot see,
Aiding you to bear your burdens,
  Heavy, painful, though they be.

Courage, darling, we are with you,
  Father, mother, all are here,
Sailing with you o'er life's ocean —
  We will guard you, do not fear.

THE ROCK IN WHICH I TRUST.

I TRUST in the Rock of Eternal Truth,
Which firm shall endure forever and aye;
When hoary creeds shall have crumbled away,
It will stand unmarred by Time's iron tooth.

I trust in the rocks of Kindness and Love,
And deeds that will make earth's children blest;
They will live when this casket is laid to rest,
And the spirit has passed to its home above.

On the rocks of Honor and Virtue I find
A sure foundation on which to stand
And rear a structure sublime and grand,
Which will ever defy the decay of time.

All these shall live and with radiance gleam
Along life's pathway, when time-worn creeds
Shall have paled in the light of noble deeds,
And have vanished away like some fabled dream.

SCARS.

"O MAMMA!" cried little Freddie,
  And his heart was rent in twain,
And he sobbed on her loving bosom
  And his tears flowed down like rain.
"Why, what is the trouble, darling?"
  The mother gently said.
"O mamma! Tom Jones has pounded
  Big nails in my nice new sled."

"You can draw them all out, Freddie."
  "O mamma, I did; but see,
I can't draw the holes out with them,
  They're as fast there as they can be."
"We will have it all painted over,"
  Mamma said, in soothing tone.
"But the marks are there yet, mamma,"
  Cried Freddie, when all was done.

MORAL.

The bitter words we have spoken,
  The acts of unkindness done,
The vows we have made and broken,
  Are recorded one by one;

And they leave on memory's tablets
  A stain that will always stay,
Defacing its sacred pages,
  And it will not wash away.

A wrong may be all forgiven,
  But the memory will still remain.
Oh, would that she were less faithful
  It would lessen many a pain
If the past could be all forgotten;
  But alas! the scars are fast,
And although the wounds are covered,
  The ghosts of the buried past

Like grim specters come to haunt us,
  And like hideous phantoms rise
To rob life of joy and sunshine
  And darken the brightest skies;
And although the wounds are covered
  The scars will, alas! remain,
Although to the outward seeming
  There remains no trace of pain.

IN MEMORIAM

STORMY CLIFF (MRS. M. M. B. G.).

Stormy Cliff, thy earth-task 's finished—
  Sorrows, struggles, pains, are o'er;
Waves of grief and tribulation
  O'er thy bark shall dash no more.

Thou hast found rich compensation
  For the joys on earth denied;
In the land of fadeless beauty
  Shall thy soul be satisfied.

Brothers, sisters, there shall greet thee,
  Whose true love will changeless be,
Shining brighter through the ages
  Of a blest eternity.

Though we miss thee from our circle,
  Listen to thy voice no more,
We can almost feel thy presence
  Wafted from the spirit shore.

We will keep thy memory verdant,
  Deck thy grave with sweetest flowers,
Till we gather fragrant blossoms
  With thee in the heavenly bowers.

THE OLD YEAR AND THE NEW.

1893–1894.

THE gray Old Year is dead;
His race is run and he is laid to rest;
The benisons of millions he has blest
Are showered upon his head.

The fresh New Year is born;
He sallies forth, and radiant on his brow
The star of hope beams forth, all glittering now,
Fair as the dewy morn.

Farewell, Old Year, farewell;
We miss the sheaves of ripened golden grain
Which thou hast garnered free from earthly pain.
'T is well, Old Year, 't is well.

We hail thee, bright New Year;
We crave a benediction at thy hand;
Let harmony and peace pervade our land
Throughout this new-born year.

May no dark clouds arise,
Dissensions dire nor internecine strife
Invade our borders, seek our nation's life,
But send us placid skies.

IN MEMORIAM MRS. M. L.,

WHO PASSED BEYOND FROM DAYTON, OHIO, FEBRUARY 25, 1892.

WRITTEN BY REQUEST.

WE will not call thee dead,
Although thy vacant form lies still and cold,
Peacefully rests beneath the dark, damp mold,
The enfranchised spirit fled.

Oh, glorious liberty!
Fettered no more by this encumbering clod,
No more the paths by erring mortals trod
Bring weary pain to thee.

But, oh, thy vacant chair!
Thy absence from the old accustomed place!
No more to hear thy voice and see thy face,
Is much of grief to bear.

Thy mortal presence now,
As was thy wont, earth's suffering ones to bless,
Comes not to soothe with loving tenderness;
But man can only bow,

Though burdened and bereft,
With calm submission to th' Eternal Will,
Whose purposes mankind can but fulfill,
Though hearts with grief are cleft.

Thy kindly words of cheer,
The voice of sympathy, the gentle tone,
Come not to uplift and guide the fallen one,
And light the pathway drear;

And yet we feel thee near,
And know that from the radiant spheres above
Thy earnest spirit comes, with tender love,
To comfort and to cheer.

Friends, let us cease to grieve
For one who 's passed beyond life's setting sun,
Life's burdens borne, life's work so nobly done,
Blest guerdon to receive.

A few more fleeting years
Will tell life's tale — and on the elysian shore
We, too, shall walk, to sunder nevermore,
Where fall no mourner's tears.

KITTY ADAIR.

THERE lived a maiden charming and fair
In the pleasant village of Lintondare,
With cheeks like roses, and golden hair,
A heart all guileless and free from care—
That beautiful maiden was Kitty Adair,
With manner and bearing most debonair,
With temper gentle and virtues rare.
But Kitty was caught in a fatal snare,
For she married a miserly, cross old bear
Who was charmed by her beauty and talents rare,
Because he was called a millionaire
And had, it was said, a bountiful share
Of stock in bank in the city of Claire.
He shut her up in his dingy lair,
Where she drooped and pined for her native air,
And gazed about with a vacant stare,
And faded away like a vision fair.

Now, innocent maidens all, beware!
Oh! be not dazzled by glitter and glare,
And heed not the world's deceitful blare,
Lest your hasty actions your heartstrings tear;
And offer to Heaven a fervent prayer
That you fall not into poor Kitty's snare.

INHARMONY.

WHAT enters an unbidden guest,
Awakes sad thoughts within the breast,
Fills the whole being with unrest?
Inharmony.

What makes life's burdens hard to bear,
Makes bitter every toil and care,
And brings of grief a double share?
Inharmony.

What fades the cheek and dims the eye,
Bids peace and sweet contentment fly,
As wearily the days go by?
Inharmony.

The heart grows sick, and sighs and tears
Fill up the measure of the years;
This hated specter still appears—
Inharmony.

Down in the soul's serenest deeps
This lurking foe insidious creeps,
And vigils at the hearthstone keeps—
Inharmony.

Within the court, without the gates,
In happy or in adverse fates,
This vampire at the threshold waits—
Inharmony.

What makes the spirit long to soar
Above, beyond this mundane shore,
Where mortal griefs intrude no more?
Inharmony.

MOTHER 'S GROWING OLD.

MOTHER is growing old, dear;
  The form once strong and fair
Is yielding to the weight of years
  And bowed by many a care;
The rose has faded from her cheek;
  Her locks of sunny gold
Are frosted o'er with silver threads,
  For mother 's growing old.

She led your infant feet, dear,
  With faithful, loving care,
And guarded youth's uncertain steps
  From many a lurking snare.
Be now her stay and succor as
  She nears life's sunset gold,
And make to mother some return
  When she is growing old.

Oh, do not cloud her sky, dear,
  By word or look unkind,
But let her sacrificing love
  Be in your heart enshrined!
Oh, fill her life with sweet content,
  All joy that heart can hold!

Encircle her with Love's strong arms,
  For mother 's growing old.

Oh, strew life's downhill side, dear,
  With fairest, sweetest flowers!
Let gladness bless her waning years
  And cheer life's wintry hours.
Each kindly act to her will be
  More precious far than gold,
And fill dear mother's heart with joy
  When she is growing old.

These patient, loving hands, dear,
  Are no more swift and strong,
For they are growing weary now—
  They 've borne life's burdens long.
A few brief years and her dear form
  Must press the dark, damp mold,
And home will hold a vacant chair,
  For mother 's growing old.

TO MRS. M. F. J.,

ON RECEIPT OF AN EASTER CARD.

THANKS, kind friend, for friendship's tokens!
  How they back my spirit bear
To the days agone and vanished,
  When we all were free from care!
On the sacred wall of memory,
  Garnered there with tender care,
Hangs full many a beauteous picture—
  Gems of friendship, rich and rare.

How I prize these cherished treasures
  As I scan them day by day,
And they shine with brighter luster
  As the years glide on their way.
Still may friendship's cords grow stronger,
  And may added radiance be
Given to all her glowing pictures,
  In the bright time yet to be.

BEAR THY BURDENS.

BEAR thy burdens, child of sorrow;
They will bear thee to thy rest,
Bring thee dawn of some bright morrow,
Where no thorns will pierce thy breast.

Thorns of grief and sore temptation
Pierce thy bosom here below;
Sweet will be thy compensation
When no pain thy heart shall know.

Anguish here may vex thy spirit;
Grief which seems too deep to bear
But prepares thee to inherit
Blessing in a realm more fair.

Somehow wrongs will all be righted;
Sometime peace will guerdon be;
Somewhere lives by sorrow blighted
Dawn in sweetest harmony;

Tears, the heritage of mortals,
Somewhere will have ceased to flow;
Zephyrs from the heavenly portals
Bear no wailing sounds of woe.

DEATH OF AN INFANT.

Farewell, sweet babe so dearly loved and cherished;
Around thy vacant form we weeping stand
And realize that earthly hopes have perished,
Torn from our grasp by death's relentless hand.

But as we look beyond with spirit vision,
An angel presence greets the wondering sight;
Free from disease, in that blest home elysian
He dwells with loved ones in the realms of light.

Secure from earth's alluring snares,—they never
Will tempt his youthful feet to turn aside
From Wisdom's pleasant paths,—safe, safe forever,
An angel form, a spirit glorified.

Oh, precious thought! that we shall meet to sever
Oh, nevermore, in the blest "By and By";
Each broken link be bound to sunder never
In the bright land where loved ones never die!

STAND BY THE FENCE.

"O MAMMA! just look at those chickens!"
  "Sh'! Sh'! Run, dear, drive them away!
They 're scratching out all those choice flower seeds
  We planted so nicely to-day.
I declare, it is too aggravating!
  I think, if we make a pretense
To have a nice lawn and grow flowers,
  We certainly *must* have a fence.

"Only yesterday, while we were dining,
  I looked through the window and spied
Neighbor Jones's old cow in my rose bed,
  And biting my shrubs off, beside.
She spoiled my clematis and *deutzia*,—
  I think it would show our good sense,
If we would avert such disasters
  By having a time-honored fence.

"Look! Brown's horse is out of the stable!
  Oh, the vases! he 's running this way!
There! he 's broken them all into fragments!
  Oh, what next will happen to-day!
They may call me a crank or old fogy—
  I think it shows wisdom and sense

To keep all unwelcome intruders
  Outside, by the time-honored fence.

"I don't mean the cumbrous enclosure
  Of boards, just as high as your head,
Nor even the nice fence of palings,
  And painted white, green, pink, or red;
But the fence that I want for *my* dooryard
  Is the iron fence, dainty and neat,
Which keeps out all troublesome bipeds,
  And sends cattle elsewhere to eat."

WRECKED.

When ships are wrecked on the treacherous main,
Their burdens of living and loving freight
In a few brief hours of suspense and pain
Are standing beyond at the City's gate,

And never a ripple or sign is left
To mark the spot on the ocean's breast
Where the tomb in the darkened water's cleft
Is hid by the billow's foaming crest;

But the *lives* that are wrecked must struggle still,
Though vainly wooing the angel Death;
No friendly waters their pulses chill,
That the weary form may yield up its breath.

They must face the world with a happy smile
While the pent-up anguish that lurks within
Is crushing and bursting the heart the while
As perchance it looks back on the "might have been."

They must cover the wounds and the torturing pain,
Hide even from loved ones the bitter grief,
Though tears in secret may flow like rain,
Till death brings kindly and sweet relief.

IS THE WORLD GROWING BETTER?

TELL me, does the world grow better?
Tell me, scientists and savants;
Tell me, philanthropic lovers;
Tell me whether crime and misery,
Vice in all its blackest colors,
Greed, and love of lucre only,
Love of pelf and power for evil,
Are a synouym for progress?

Wretchedness and crime are rampant;
Right and honor count for nothing;
Truth 's a bird of rarest plumage;
Purity is held at discount;
Virtue is a cast-off garment;
Justice has deserted mortals,
Plumed her wings and flown to heaven.

City, village, town, and hamlet
Reek with every foul pollution,—
Treachery in all high places,
Striving for the spoils of office,
Seeking each his own, regardless
Of the happiness of others,
Strife of man to crush his fellows,

Spurning right and trampling justice,
Crushing out the pure and noble.

And I fear that countless eons
Will have run their onward courses,
Will have passed and be forgotten;
Varied change and revolution
(Heaven ordain it may be bloodless!)
Shall go thundering down the ages
Ere friend Bellamy's prediction
Will be realized by mortals.

Mayhap each diseased condition,
Howsoever foul and fetid,
Holds within a latent healing.
Is there no inherent process
Lurking in fair Nature's forces
By which vice and crime are sated,
Working thus their own decadence,
Compassing their dissolution,
Perishing by their own corruption,
By survival of the fittest?

If these omens are the signals
Of a coming revolution,
Of the new and radiant dawning
Of the morn of man's redemption
From the tyranny of evil,
From the chains of lust and passion,

Then time may be drawing nearer,
Then an era be approaching,
When our planet may, through changes,
Be at last regenerated,—
Enter into new existence;
Peace and plenty crown the nations;
When from crime, and want, and squalor,
Penury, and destitution,
Bitter hate, and blood, and carnage,
Strife, and internecine warfare
Earth shall be emancipated,
And the wrongs she erst has suffered
May be reckoned by the people
Memories of barbarous ages.

Woman's epoch is approaching;
She shall be a serf no longer;
She will rise and claim her birthright,
Freedom's boon bestowed by heaven.
She will be no longer subject
To the proud "lords of creation";
Will not be adjudged and punished,
Punished by her lordly brothers
For a crime she has committed,
Held by them in vilest durance,—
Aye, led forth to execution,—
By the statutes of a nation
Which she has no voice in framing.

Woman has for untold ages
Been the bond-slave of her brother,
Been a serf to do his bidding,
Been to him a mere appendage.
Time is dawning when, an equal,
As a peer in any station,
She, with dignity befitting
Queen or duchess, stands beside him.
Then shall come a reign of Reason,
Banishing the present evil.

Womanhood will hail the dawning
Of the coming morn of freedom,
Of the bursting of her fetters,
Breaking of the chains that bind her,—
Hail with joy the golden era
Of her full emancipation.

Many here may not behold it
With the eyes of earth-born mortals;
But from out the home elysian
We may gaze with holy rapture,
Feel the joy that fills the bosoms
Of the dear ones left behind us,
Thrilling with the bliss of freedom,
Borne aloft on eagle pinions,
Feeling life is worth the living.

TO A BLIND LADY ON HER NINETIETH BIRTHDAY.

STILL adown life's turbid river,
Rolling on with ceaseless flow,
Onward to the boundless ocean,
Glides a barque launched long ago.

Clouds have come and storms have gathered,
Wild winds blown and waves dashed high,
Still the boat bears bravely onward,
Bidding fear and danger fly.

Calm and peaceful be thy voyage,
Drawing nearer to its close;
Nearer to the port of safety,
Where earth's weary ones repose.

There shall cease thy toils and conflicts,
Carping care and weary pain,
Sorrows end in peace and blessing,
Joys in full fruition reign.

Sightless orbs, now veiled in darkness,
Then shall gaze on visions bright,

And the soul's now shaded windows
  Be illumed with heavenly light.

While the evening shadows gather,
  May new radiance cheer thy way,
Making all the entrance golden
  To the glad immortal day.

THE MOTHER'S LESSON.

A CULTURED mother of thoughtful mien,
With pleasing visage and brow serene,
Was blessed with a bevy of bright young girls,
With silken lashes and sunny curls.

There was black-eyed Bessie, and blue-eyed Nell,
With Kitty, and fairy Annabell;
And one of the mother's most fervent prayers
Was, "God keep my darlings from fashion's snares."

"I want them," she said, "to grow strong and hale,
With cheeks like roses, not wan and pale.
I can keep them all right while under my eye,
But they 'll grow individuals by and by,

"And all the young ladies nowadays
Compress their chests by corsets and stays,
And it pains my heart to think that mine
May follow soon in the beaten line.

"I will teach them — that 's all that a mother can do —
The lessons of Nature sublime and true,
And trust to their reason and sense the while
To shun the paths of folly and guile."

And, reasoning thus, she closed her door
And wended her way to a jewelry store,
Where she ordered a clock case made close and small,
But with large, strong works and hands and all
To be pressed in fast so it could not run,
And sent to her home when the work was done.

"It can be of no service," the jeweler said;
"It won't keep time," and he shook his head.
"Never mind," said she, with a wistful eye,
"I will have my use for it by and by."

"He thinks me a fool or a crank," said she;
"That name has lost all its terrors for me;
So they said of Galileo when he knew
That the earth revolved, and proved it true."

The work was finished and soon sent 'round—
The oddest clock that ever was found.
It was flat, and narrow, and straight, and tall,
And the children questioned her, one and all:
"Mamma, what do you want with that pent-up thing?
Why, the pendulum has no room to swing."

"I want it, dearies, and sometime you
Shall know what use I shall put it to."
So she stowed it away, and the mother thought
That soon by the children 't would be forgot;

But she knew that the time ere long would come
When her earnest lesson must be brought home.

So time ran along at lessons and play,
Till Bessie came home after school one day,
Saying: "Mamma, can't we go shopping uptown?
I want a corset like Mamie Brown."

"A corset, darling? Why, what for, Bess?
Are you not quite warm in your new wool dress
And your nice, soft wrap, so snug and warm?
And your gossamer keeps out the rain and storm."

"Oh, I 'm not cold, mamma! but Mamie Brown
Says my waist 's too big,—I must lace it down,—
And I 'll grow ill-shaped, like Sally Gray,
If I don't wear a corset every day."

"What fault can you find with Sally Gray,
With her pleasant voice and her winsome way,
Her cheeks that vie with the roses bright,
And her eyes that shine like the stars of night?"

"Oh, mamma, I know she 's kind and good,
But she don't look neat like *young ladies* should!
Why, her waist is growing so big and round
That you 'd think she weighed two hundred pound."

"Young ladies! I thought they were little girls,"
The mother said, as she stroked their curls;
And she went to the closet, where, stowed away,
The clock she had placed there forgotten lay.

"Now, children, why is it this clock won't run?"
"Why, it has no room," answered every one.
"It is shut up so close that the wheels can't go.
What good is it, mamma, I'd like to know?"

"It is *no* good, darlings, unless it preach
To your hearts the lessons I long to teach;
But should it to your reason these truths unfold,
I shall value it more than a mine of gold.

"Now you see, my dears, that a clock can't run
Without plenty of room for the wheels to turn;
And the pendulum must have a space to go
Backward and forward, steady and slow.

"Now, we'll liken your chest to this clock, my dears,
Which to all your eyes of no use appears.
All these little wheels your lungs shall be,
Which to breathe pure air must be loose and free.

"And your stomach and liver the springs we'll call,
And the hammer, and wires, and hands, and all
Are like to your nerves and veins, which bear
A part in this clock so odd and rare.

"Your throbbing heart, which beats on and on,
And never can cease till your life is done,
Shall be the pendulum swinging slow
Backward and forth as the moments go.

"Now you see that machinery cannot turn
Without a space for the works to run;
And do you expect the form divine,
Of delicate fabric and texture fine,

"To live, and expand, and move, and feel
In a shut-up cage of bone and steel,
Where your lungs are like these wheels compressed,
And close imprisoned within your chest,

"And your heart, like this pendulum, pent up so,
Can't do its work, and expand, and grow,
Your stomach and liver cannot be free
To perform their functions? And don't you see
That disease, with its pallid touch, will come,
And health depart and its rosy bloom?"

The children listened and pondered well,
And the mother hoped that the good seed fell
Into fertile soil and would all take root,
And yield a harvest of precious fruit.

The family all went to church one day
To hear the minister preach and pray,

And list to the organ's mellow notes
As the music pealed from its deep-toned throats.

In the pew before them demurely sat
A tight-laced lady, with gay, plumed hat.
"Look, mamma!" said Kitty, the dear little thing,
"The pendulum has no room to swing."

HEART WOUNDS.

THERE is something often to wound the heart,
Something to make the hot tears start,—
A thoughtless action, a careless word,
And the soul to its nether depths is stirred;
But the blessed tears relieve the pain,
And we take up the burden of life again.

There is something often to mar our joy—
Rarely is gold found without alloy;
Bitterness always is mixed with sweet,
The dearest pleasures are those most fleet,
The fairest roses the soonest fade,
The brightest hopes in the dust are laid.

Perhaps it is well: the frost and snow,
The dashing rains, and the winds that blow,
As well as the sun's all quick'ning beams,
When he floods the landscape with golden streams,
Prepare the earth that a fruitful soil
May give rich reward for the lab'rer's toil.

It is thus that tears and trials come
To prepare the soul for a brighter home;
To purify, chasten, subdue, refine,

That the gold within us with luster may shine,
That the spirit may come forth pure and bright,
All clad in the garments of truth and light.

There may come a time when grief and tears
(If not in this, in the holier spheres)
In joy's pure sunlight will melt away
And love's sweet power every breast shall sway;
When the rose without the thorns may bloom,
Nor life be shadowed by sorrow's gloom.

A LETTER IN RHYME.

TO MISS B. E. S., DELAWARE, OHIO, MONNETT HALL.

DEAR BERTHA:—

How I would enjoy,
  This bright December eve,
To peep into your cozy room;
  'T would be a kind reprieve

From dull monotony and care,
  And drive the blues away;
But this can't be, for you to-night
  Are many miles away.

Bertha, I missed the roses so
  Last fall, to send to you;
It was so dry they could not bloom,—
  I know you missed them, too.

And I have hoped next year to see
  Their glorious blooms appear,
But now our weather prophet says
  We 'll have a drouth next year.

Well, how the months do speed away!
  Christmas will soon be here,

When we may see your face again,
  Just for a while, my dear,

And hear you talk of college life;
  Oh, how I would, I ween,
Enjoy to be your college mate,
  If I were sweet sixteen!

My school days, although passed and gone,
  Are still to memory dear,
And on its pages many names
  Of school friends still appear;

And oh! how pleasant yet to me
  Those moments to recall,
The brightest of my girlhood's years,
  The happiest of them all.

Aunt Lizzie 's gone to see your ma;
  Went home with Carl, last night;
And at the threshold she will miss
  Our Bertha's greeting bright.

Well I must stop this talk, or I
  Will make you sick for home;
Soon Christmas will be here, and then
  We hope to see you come.

La grippe is all around again,
  And only yestermorn

A funeral took place in town,
  And Grandpa B—— is gone.

I hope the scourge won't search *you* out,
  We want you home again.
Good-night, my dear, and happy dreams.
  Yours lovingly,
                                        Aunt Mane.

RETROSPECT.

INSCRIBED TO MRS. S. J. E.

ANOTHER year, my dear friend, has flown,
  Swallowed up in the boundless eternity;
And our feet still press these mundane sands,
  Our barque is yet tossed on life's billowy sea.

How many have passed to the unseen shore,
  And their tenantless forms lie cold and still;
They have entered the grander life beyond,
  And others their vacant places fill.

The year that is dead, and is gone for aye,
  To some has brought pleasure and calm delight,
While others have drunk of the cup of woe,
  And morning has vanished in deepest night.

How many in billows of smoke and flame,
  In agony bitter, and anguish dire,
While naught but the blackened forms remained,
  Have yielded their lives to the demon fire!

'Mid the foaming billows mad and wild,
  In the shadowy depths of the ocean wave,

How many have sunk 'neath the waters cold,
  And their forms uncoffined have found a grave!

Destruction and death have been left behind
  In the storm king's devastated path;
The homes laid waste by his ruthless hand
  He has scattered wide in his furious wrath.

Farewell, Old Year; to the stern behest
  Of Infinite Power we all must bow;
Our paths are marked to life's furthest bound,
  And Destiny's seal is on every brow.

He holds us all in his iron grasp;
  Along each pathway lurks seeming ill;
But the end is good, let us calmly wait,—
  "Each cloud has a silver lining" still.

It must all be best,—the hope deferred,
  The wish unfulfilled, the vanished dream,—
In the clearer light we shall sometime see
  That the shadows obscured the sunlight's gleam.

And sometime, my friend, when the New Year comes,
  We'll be anchored safe on the further shore,
Where the storm-tossed mariner rests in peace,
  And the wild waves above him shall dash no more.

A VOICE FROM THE SUMMER-LAND.

[Dedicated to Ira J. and Lue R. Powell, on the occasion of the translation of their little daughter, MABEL ALDA, who passed from the mortal tenement, August 5, 1884, near Urbana, Champaign County, Ohio, aged six years and six days.]

YOUR Mabel's safe in heaven, mamma,
From pain and suffering free,
And happy children throng around,
Their new-born friend to see;
And grandma took me by the hand,
And led me to her home;
She looks so pleased and happy now
Her little Mabel's come.

You must not weep for me, mamma,
For that will make me sad;
We 'll come to see you every day,
And try to make you glad.
They say I am an angel now,
But I 'm your darling yet;
The loved ones of my earthly home
I never will forget.

You must not think I 'm dead, mamma,
Because I 've gone away;

My body was worn out with pain,
  I could not longer stay;
And grandma says she 's wanted me,
  She is so glad I 've come,
And you have sister Gertrude left
  To cheer your earthly home.

This land is fair and bright, mamma,
  The birds so sweetly sing,
And flowers are blooming everywhere,
  And sweet as breath of spring;
I know I will be happy here,
  With friends so true and kind,
And little hands, the angels say,
  Shall sweet employment find.

I bring my love and kiss, mamma,
  To those I 've left behind,
To papa, sister, auntie dear,
  And all my friends so kind;
Sometime you all shall come to me,
  And in this home so bright
We 'll all be happy evermore.
  Now, dear ones all, good-night.

THE NEEDS OF THE AGE.

Written in Answer to One of My Critics.

The world is, of course, growing better,
  But heaven knows it 's bad enough yet;
There 's room for a wondrous improvement —
  Just now we 've no virtues to let;
'T is true we 've progressed, but progression
  Is found very needful to-day,
And though we 've outlived the dark ages,
  Might still over reason holds sway.

When woman has purer conditions,
  The race greater, nobler, will grow;
She grandly will fill her high mission,
  And on future being bestow
The boon of sublimer existence,
  A spirit exalted and free;
No spot on the physical structure,
  But tuned all to sweet harmony.

When man 's not enslaved by tobacco,
  When rum never maddens his brain,
When intellect yields not to passion,
  And will has full power to restrain,
Then home will not be but a shelter,

A refuge from tempest and storm,
But frescoed with Love's sweet adornment,
And guarded by hearts true and warm.

When follies and foibles of fashion
Man ceases to love and admire,
Soon woman will banish and shun them,
And soon of their slavery tire;
For his eyes she laces and ruffles,
For his eyes she powders and paints;
If man would denounce and condemn them,
He 'd soon have no cause for complaints.

And while *man* demands social vices
There always will be a supply;
When fathers become pure and noble
And true as the stars in the sky,
Mankind will arise and look upward,
Stand forth clothed with honor and truth,
The earth will don garments of gladness
And joy displace sadness and ruth.

Not till woman has purer conditions
Will vices and crimes ever cease;
Not till she exults in her freedom
And homes of affection and peace,
Will earth bear the image of heaven
And virtue, pure gem from the skies,
Irradiate earth with her brightness
And light from the shadows arise.

WEARY.

THIS life would be joyless and dreary
  And its burdens too heavy to bear,
With its griefs and its sad disappointments
  And its measure of sorrow and care,

With its sweetness to bitterness turning,
  Its cups dashed untasted away,
And the hopes we so fondly have cherished
  Doomed only to blight and decay,

Were it not for the cheering assurance
  That the sunshine comes after the rain,
That beyond there 's a sweet compensation
  For all the heart's anguish and pain.

We will count there as naught all earth's sorrows,
  Earth's loss we will there count as gain;
All hail to that glorious morning,
  Heaven's peace for earth's conflicts and pain.

FOR MY COUSIN'S ALBUM.

IN the after years, dear Emma,
  In the sometime far away,
When your cheek has lost its freshness
  And your locks are tinged with gray,
When your children stand around you
  In the strength of manhood's pride,
Strong in lofty aim and purpose
  As adown life's stream they glide,

You may chance to turn these pages,
  And this earnest wish I 'll trace,
That through all life's weary journey
  As the years have flown apace
You 'll have walked with strength unfaltering,
  Nobly wrought in word and deed,
Have fought bravely life's great battles,
  And may claim the victor's mead.

IT ALL FLIES OUT AT THE WINDOW.

"I WILL marry for gold," said a maiden fair,
As she tossed her ringlets of raven hair,
"For *Love* is a bird of plumage rare
Which will soon fly out at the window."

She wedded for wealth and the power it brings;
It glittered awhile, but it soon took wings
And followed the wake of all transient things,
And it all flew out at the window.

"I will marry for fame," another said;
"He must have renown, the man I wed.
'T will live and endure when gold has fled,
And will not fly out at the window."

So she wedded for fame, this maiden sweet.
For a time the nations sat at her feet;
But just as she reckoned her bliss complete,
It all flew out at the window.

Then a sweet girl said: "True love will be
A beacon light o'er life's surging sea.
I'll wed for love, it will stay with me,
And will not fly out at the window."

"She married for love; and its
rosy light
Made earth an Eden, all fair
and bright."

She married for love; and its rosy light
Made earth an Eden, all fair and bright;
But time brings change, and it took its flight
And vanished out at the window.

Whatever, dear girl, may be your fate,
Whatever your portion in wedded state,
Unless you are joined to your soul's true mate
Love, too, will fly out at the window.

THOUGHTS AT AN INEBRIATE'S BURIAL.

Another barque has crossed the silent river,
Has gained the further shore,
Has launched into the limitless forever,
The mortal conflict o'er.

Without volition cast upon life's ocean
By Fate's relentless hand,
A prey to every base and vile emotion,
No strength given to withstand

The appetites which lured him to perdition,
Assailed on every side,
Weak and unstable, born to low condition,
No power to turn aside.

O friends! spread charity's broad mantle over
His faults, whate'er they be;
Let sweet forgiveness, like a garment, cover
Each frailty we may see.

The soul that here was swayed by human passion
May have a purer birth;
May come forth clothed in a diviner fashion
Than Fate ordained on earth.

May bitter feelings from our hearts be driven,
  Above this helpless clay.
Forgive, as we will wish to be forgiven
  When comes our trial day.

Birth and environment make stern conditions
  Which mortals cannot flee;
We wait with patient trust the soul's transition
  To make us wholly free.

THE POWER OF KINDNESS.

GENTLE words are full of sweetness;
  Each a healing balm imparts,
Dropping like the dews of evening
  Into crushed and bleeding hearts;
As the showers that fall from heaven
  Make fair nature to rejoice,
Gentle words refresh the spirit,—
  Tender tone and quiet voice.

Husband, when upon returning
  From the labors of the day,
If your wife should fail to greet you
  In her old accustomed way,
Don't proceed to scold and chide her
  Or imagine she is ired,
But just say, in kindest accents,
  "Dear, I'm sure you must be tired."

Wife, if husband comes home weary
  And perplexed with business care,
If he's cross and falls to scolding
  And you chance to get a share,
You may stifle the volcano
  If you can but gently say,

Just before it bursts upon you,
"Husband, are you sick to-day?"

Parents, when you have occasion
(As good parents often do)
To reprove a wrongful action
And advise the good and true,
Do it gently: words of kindness
Always melt and move the heart;
Anger does not conquer evil
Or inspire a nobler part.

If your friend should chance to wound you,
With perhaps no wrong intent
(Thoughtless words are often spoken),
Be not hasty to resent.
Do instead some act of kindness,
Though with grief your heart should burn;
Pour oil on the troubled waters;
Do not wound him in return.

Friendship thus will be cemented,
Wrought into a golden chain—
Bond which time nor death can sever,
Which enduring will remain.

DEATH THE BEST FRIEND.

O BLESSED Death! Thou precious, priceless boon,
Thou truest, dearest, ever-constant friend
To sorrowing mortals; thou didst never fail,
While other friends have proved, alas! untrue.
Thou givest to the weary pilgrim rest,
Food to the famishing; at thy pure fount
Earth's thirsty children all may drink and live
For evermore. Thy soothing kiss, when pressed
Upon the lips of helpless sorrow,—joy
Turned all to gall and wormwood, solitude
Too deep for words, the solitude of soul,
Life all embittered by sense of wrong,
The roses plucked, naught left but cruel thorns,—
Is sweeter than the nectar of the gods.

O beauteous angel, Death! Thou dost give back
Unto our fond embrace our loved and lost,
The cherished ones whose forms we long ago,
With sad farewells and bitter, burning tears,
Laid tenderly away beneath the ground
And strewed the sacred mound with fragrant flowers.
How sweet unto the weak and tott'ring form,
With locks all whitened by the frosts of time

And bent by weight of years and weariness,
Is thy caress.

Thou givest back our youth,
Our strength and vigor, comeliness, and all
That makes existence beautiful and sweet.
Men call thee cruel; nay, thou art most kind;
All that we lose in life we find in thee.

CAN'T GIVE UP THE ROSES.

SAID Farmer John
to his wife one
day:
"You better dig out
them posies;
We need that spot for a cabbage bed;
'T will pay more than all your rosies.

"It 's queer to me, and I can't see
why
All women will fuss and flurry
'Bout what don't 'mount to a row of pins
And makes only work and worry."

"John, don't you know," the good wife
said,
"That the Good Book plainly teaches
That man cannot live by bread alone?
He needs some grapes and peaches.

"We need potatoes, and wheat, and corn,
And the fresh, sweet clover posies,
The ripe tomatoes, and apples sweet,
But we can't give up the rosies."

TO A YOUNG FRIEND.

My youthful friend, now in life's radiant morn,
When skies are all aglow with rosy light,
And gilded visions dance before thine eyes,
Set careful watch and guard upon thy life;
For know that every wrongful act will leave
A stain *indelible* upon thy soul.
After the wound is healed, the scar remains;
The storm fiend leaves his devastated track;
So every wrongful action of our lives,
Howe'er regretted or repented of,
Will mar the spirit by its darkening shade.
Oh! if mankind but fully realized
This, Nature's holy truth—a truth inscribed
Upon our being's page, and graven there
In characters of fire—and all could feel
That every sin will stain and scar the soul,
That life's young tree, rooted in error's soil,
Will grow deformed, unsightly, void of grace,
And harvest yield of tasteless, bitter fruit,
How 't would restrain from wrong; and there would be
Less grief in human hearts, and more of joy;
Peace, like a benediction all divine,
Would find a home with all beneath the skies,
And earth would be a happier dwelling place.

SORROW'S RECOMPENSE.

A Vision.

"Hath sorrow recompense?" I asked my heart;
"What compensation for our direst woe
Is here vouchsafed? Is this short life a part
Of Hades, and has stern Fate willed it so?"

"Relentless, hard, inevitable Fate,
Thy stern decrees spare not; the small, the great,
Must bow before this despot—at his nod
Comes bale or dole, comes blessing or the rod.
Ye Powers Above, have pity! Is it true?"
My tortured spirit cried. Then straitway through
Ethereal void appeared an angel face,
And smiled into my soul with kind, sweet grace,
Poured in my wounded heart a healing balm,
Filled all my being with a holy calm.
I gazed with rapture on the radiant face.
She drew me to her in a close embrace,
Then bore me hence above the star-gemmed skies,
Beyond the rose-wreathed gates of paradise.

"Behold, sad child of earth," the vision said,
As through the vine-clad bowers she gently led
And bade me rest beneath their fragrant shade.

I stood in wonder, as she softly laid
Her hand upon my head and said: "Behold
The glories which I here to thee unfold—
Joys of which mortals never knew or dreamed
[Her face with love and heavenly sweetness beamed].
Thine eyes on earth are veiled, thou canst not see
The possibilities in store for thee;
Wait patiently, dear child; thine aftermath,
When garnered what delicious odors hath;
Look out upon the grandeur of these spheres,
See compensation for thy pains and tears;
All this is thine thy talents to employ,
Thine for eternal ages—thine the joy
Of boundless opportunity—thy mind
Untrammeled shall new springs of knowledge find.
Eternal progress is the guerdon here
For ruth and anguish in the lower sphere;
Thy faculties, which erst imprisoned lay
Within a cumbrous tenement of clay,
Unfettered here may soar with airy wings
To heights unscaled by time and sentient things.
Afar beyond the ills of time and sense
Find here thy sorrow's glorious recompense."

The heavenly vision vanished—in my room
She left the fragrance of the roses' bloom,
And in my soul a sweet and restful peace
(Oh, could it linger till this life's surcease!)
Which made me strong to suffer and endure,
Well knowing that my recompense is sure.

TO MISS E. L. S.,

A MEMBER OF THE ENTRE NOUS LITERARY SOCIETY AT OSBORN, OHIO.

You are going hence, dear sister,
  From your pleasant village home;
Other duties soon will meet you,
  Other fields you soon will roam;
And, while we shall miss your presence,
  May we fondly hope that you
Will, though absent from our circle,
  Sometimes think of *Entre Nous*.

Paths untrodden lie before you—
  We could wish them strewn with flowers;
May you find sweet violets blooming
  'Long life's streams and shaded bowers;
And when sometimes friends in Osborn
  Rise before your mental view,
May we hope you 'll still in memory
  Keep a place for *Entre Nous*.

May your life's fair, radiant morning,
  As ascends its rising sun,
See with firm, unwavering purpose

Every duty nobly done;
And when other friends shall greet you
With caresses warm and true,
Will you keep some corner verdant
In your heart for *Entre Nous?*

But when, sometime in the future,
You shall somewhere chance to meet
The *one* who above all others
Will make home and life complete,
Then I fear me much that memory
Will to Osborn prove untrue,
And that in your sweet enchantment
You'll forget the *Entre Nous.*

MARAH.

O PEACE! thou gentle, white-winged dove,
  Come rest within my heart,
And bid this raven of despair
  From out my soul depart.
O angel guardians, linger near
  And still my soul's unrest,
And whisper to my spirit's ear,
  "Whatever is, is best."

The gods' mills slowly grind, but grind
  Exceeding small, 't is said,
And retribution never fails
  To fall upon the head
Of him who sins, and every wrong
  Will somewhere be redressed.
Be still, my heart, and only trust,
  "Whatever is, is best."

Though on the desert sands of life
  Our weary feet must tread,
A green oasis lies beyond
  And rest is just ahead.
Our paths through vales of myst'ry lie;
  The storms on ocean's breast
The good ship oft to the harbor bring;
  "Whatever is, is best."

MUSINGS.

If the flowers in earth's dark valley
  Grow so wondrous fair and bright,
And their graceful, fragrant beauty
  Thrills the being with delight,
What will be their peerless radiance
  When upon the shining shore
We shall gather them with loved ones
  In the blest for evermore?

If the golden chain of friendship
  Binds each heart so closely here,
If within each soul its echoes
  Are on earth so sweet and dear,
What new joys will fill the spirit,
  And what ravishing delight,
When friends meet to sunder never
  In the glorious land of light?

If when souls are truly mated
  Earthly bliss is deemed complete,
If when love fills all the being
  Pain and bitterness are sweet,
What shall be the dear enchantment
  When beyond Time's surging tide

Souls congenial meet and mingle
  By the crystal river's side?

If in earth's harmonious dwellings
  So much happiness is found,
If where homes are pure and restful
  Peace and joy so much abound,
Oh, how fond will be the blending
  On the everlasting shore,
When the cords that here were sundered
  Are united evermore!

ANSWER TO HER LETTER.

DEAR GERTRUDE:—

Oh, no! I did not write
A poem upon the Wheel;[1]
My Muse would never consent to ride
Within such a frame of steel.
Besides, I 'd have missed the glorious view
Of all the white world below,
Failed also to note how near to heaven
The sinners of earth can go.

On *terra firma* I must remain,
At least till my pinions grow.
Imagination's aërial flights
Are all that are mine to know.
They bear me up through the trackless void
To the realms of endless space,
And I visit in fancy the starry spheres,
Their beauty and grandeur trace.

I visit Saturn, with all her moons
And her rings of shining gold,
And gaze on her skies at the twilight hour,
Which such radiant views unfold.

[1] Ferris Wheel.

I even float up to the peaceful home
  Of those whom we call the dead.
I dream that St. Peter has let me in;
  I wake, and my dream has fled.

Your letter, my dear, was very sweet.
  Won't you write me soon again?
I wish you a merry Christmas time.
  Yours ever, with love,

AUNT MANE.

AFTER A WHILE.

After a while, if we patiently wait,
We shall enter in through the vine-wreathed gate,
Where the strife and turmoil of life shall cease
And the weary spirit shall rest in peace—
After a while.

After a while, on the beautiful shore,
When the weary conflict of life is o'er,
Joyfully sweet will the welcome be,
Grateful our peans of victory—
After a while.

After a while, if faithful and true
To the labor given us on earth to do,
The battle will cease, the warfare be o'er,
The guerdon be sweet on the shining shore—
After a while

After a while, if to human needs
Our hearts have responded in noble deeds,
If when we were blessed with bountiful store
Our hands have been open to feed the poor,
If we 've poured balm in each stricken heart
And the demon despair have bid depart,

Though the sowing seemed but withered leaves
With joy we 'll garner the precious sheaves —
After a while.

Each kindly action, each word of love,
Will deck with beauty our house above;
Thus, day by day, whatever betide,
We 're building our home on the other side —
After a while.

After a while — cease, my heart, to repine!
Out of thy darkness a rainbow shall shine,
Shedding its radiance and joy on thy path,
Aiding to gather thy sweet aftermath —
After a while.

After a while this encumbering clay
Shall be laid peacefully, gently, away;
Free and unfettered the spirit shall soar,
New scenes of beauty to seek and explore —
After a while.

FRAGMENTS.

Submission.

Oh, for a calm, abiding trust
  To dwell within my breast,
In sweet submissiveness to feel
  That all that comes is best!
Be patient, O rebellious heart,
  Against unsparing Fate;
'T is vain to strive, and vain to pine;
  Be still, and calmly wait.

To a Book of Poems.

O casket rare! from which Truth's radiant beams
Shine forth from every pure and precious gem,
Thou art a star, whose light serenely gleams
In silvery brightness from Heaven's diadem.

Spring Roses.

What sweet inspirations, what freshness, they bring,
These dear, speaking beauties, the roses of spring!
The gentle breeze scatters their odors around
Like heart of fond maiden with love's sweetness crowned.

To a Young Friend on the Gift of a Book of Poems.

The earnest wish, my dear young friend,
That on this page I trace
May meet thy glance some future day,
When years have flown apace.

May destiny be kind to thee
And strew thy path with roses;
May love's kind hand each thorn remove,
Until life's drama closes.

Riding in the Ferris Wheel.

Up in the Ferris Wheel we go,
Like an eagle we soar on high,
Leaving the "Midway" far below,
Sailing up to the clear, blue sky.
Up in the ocean of air and light,
Like a ship on the boundless sea,
Under our feet is the City White —
Nearer to heaven than we'll ever be.

Not to Her.

Not to her whose feet have wandered—
Who life's golden years has squandered
In the giddy maze of fashion
 Does the victor's mead belong,
But to her whose firm endeavor,
With a will which falters never,
With an earnest, noble purpose,
 Stays the tide of human wrong.

MY FRIEND OF LONG AGO.

CAN it be? I must be mistaken!
  But the face is so like her own,
With the eyes so dark and lustrous,
  And the same sweet, silvery tone.
Ah, yes, it is she! and I cover
  The face and the hair of snow
With kisses of fond affection,
  My friend of the long ago.

Our girlhood was spent together,
  And our friendship was tried and true;
We roomed together at college,
  And were comrades through and through;
But in after years we were married,
  And our lots were asunder cast,
But we held as a priceless jewel
  The memories of the past.

We had lost all trace of each other—
  In a home 'neath the Southern skies
She had seen her sons and daughters
  In virtue and strength arise;
But when the war tocsin sounded
  And its bugles began to blow,

"At length came a scene of carnage,
And her brave boys fell that day."

She was forced to yield up her treasures,
  Dear friend of the long ago.

War clouds enveloped the nation
  And covered it like a pall,
And voices from many a household
  Answered the bugle call;
The once happy home was broken,
  And husband and sons must go,
And she yielded them up to their country,
  My loved friend of long ago.

Oh, the months and years of waiting
  Full of anguished hopes and fears!
The days full of weary watching,
  Nights of fervent prayers and tears!
And when the dread news of battle
  Came flashing over the wire,
The dire suspense of the tidings
  Set heart and brain on fire.

At length came a scene of carnage,
  And her brave boys fell that day,
With face to the traitorous foeman,
  On the field of the deadly fray;
And the husband, too, lay wounded,
  And they bore him away to die,
Afar from family and kindred
  And home, with no loved one nigh.

They sleep in the sunny South-land,
  The region of blood and strife,
While the widow carried her sorrow
  And the burden of human life.
Misfortune had spoiled her substance,
  Filled her cup with grief and woe,
But she struggled on for existence,
  My loved friend of long ago.

I, too, am in solitude walking,
  And around me on mem'ry's wall
Many pictures I fain would banish
  Confront me like some dark pall;
But plenty is still my portion,—
  I 'm so grateful that this is so,—
I will shelter now and protect her,
  My comrade of long ago.

"They sleep in the sunny South-land,
The region of blood and strife."

THE ROSE.

SHE sits in her regal perfection,
Unrivaled in gardens and bowers;
I bow to her scepter, and crown her
The queen of all beautiful flowers.
I love all the sweet floral kingdom,
They all many pleasures impart;
But oh, give me roses forever!
They hold the first place in my heart.

Oh, weave her a garland of roses
Who stands at the altar to-night,
Fit emblems of love and devotion
Which sorrow and time cannot blight.
Yes, strew them all over her pathway;
In years yet to come they may be
A mem'ry to comfort and cheer her,
While sailing life's turbulent sea.

When low lies the head in the casket,
Bring roses to soften the gloom;
Their presence can light up the darkness,
And beautify even the tomb.
They point us to gardens eternal,
Where roses unfading shall blow,
Where sorrow and death do not enter,
And loved ones no parting shall know.

COMMEMORATIVE THOUGHTS.

On the 19th of August, the year '93,
We arose with the sun, and were all glad to see,
Though the weather was torrid, and dusty, and dry,
A bright, pleasant morning and clear, cloudless sky;

For the pioneer meeting, as time rolled around,
Was appointed to-day, and for this we were bound.
So, with lunch baskets packed, we set out for a while
For a drive through the country to old New Carlisle.

Soon the pioneers gathered from far and from near,
The fathers and mothers and patriarchs dear,
With children and grandchildren, all come once more
To find joyful greetings from old friends of yore.

For me this quaint village holds many a charm;
The memories that cluster are still fresh and warm,
And back to my childhood in fancy I roam
In fond recollections of mother and home.

How often have I in the dear long ago
Passed o'er Honey Creek's waters, as onward they flow,
In the old family carriage, to visit a friend
Or to the old chapel our footsteps to bend,

With father and mother and sister so dear,
Whose forms now repose in the cemetery near—
Not dead, but passed on to that beautiful goal
To live and progress while the ages shall roll.

But to me an enjoyment to-day greater still
Was to meet my loved teacher[1] of old Linden Hill,
To look in his face and behold him once more,
And hear his kind voice as I heard it of yore.

Forty years have rolled onward, and Time in his flight
Has left his stern traces; the eye's radiant light
Is dimming with age; when these signals I see
They warn me that this our last meeting may be.

Be this as it may, I shall always revere
His time-honored precepts that fell on my ear;
They were true, and ennobling, and lofty, and pure,
And will live in my memory while time shall endure.

[1] Professor Thomas Harrison.

GIVE ME THE FLOWERS NOW.

I STOOD to-day by the open grave
  Of one once so young and fair,
And roses sweet decked her cold, white brow,
  And lay in her clustering hair;
And my heart cried out, as I gazed upon
  The roses that pressed her brow,
"Don't keep them all for my cold, dead form,
  But give me the flowers now!"

Her life path was dark, and its desert sands
  Her weary, lone feet had pressed;
With sorrow and tears her heart was filled,
  By the sunshine of love unblessed;
But when death came, there were willing hands
  To place roses upon her brow;
But within my soul rose the earnest prayer,
  "Oh, give me the roses now!"

Oh, give me the flowers from friendship's tree,
  To gladden life's lonely way!
Oh, scatter their fragrance along the path
  Wherever my feet may stray!
For I cannot enjoy their sweet perfume,
  When they grace my marble brow;

I need them now to sustain and cheer—
  Oh, I want the flowers now!

Oh, give me, dear friends, the flowers of truth
  And affection, pure and sweet;
They will pave through the mountain steeps of life
  A path for my weary feet.
Don't keep them hid till the death damps come
  And gather upon my brow;
The cold, white clay cannot answer back—
  Oh, I want the flowers now!

The flowers of sympathy—how they lift
  The burdens we all must bear;
They gild and brighten each lonely hour,
  And lighten full many a care;
But the silent form does not feel their breath,
  Though entwined on the clay cold brow—
O friends, dear friends, while my heart beats warm,
  Oh, give me the flowers now!

SONNET.

THE sweetest joy of heaven, I 've fancied oft,
Will be the mingling of congenial souls.
Oh, joy unmeasured! Oh, dear, precious dream,
From which we need not wake again to feel
The disappointment of our waking hours!
Oh, bliss supreme! Oh, fullness of delight!
Each kindred spirit blending with its own,
Linked heart to heart in dear companionship,
Its own in feeling, sentiment, and thought,
In love, in sympathy, in tenderness;
Oh, rapture of delight! to know that there
Inharmony and discord have no place,
No entrance to this sacred home of peace,
And all may gather life's sweet aftermath.

THE SKELETON.

I CHANCED in her chamber to look one day,
  And there I saw revealed,
O'erhung with drapings of silk and gold,
  A skeleton's form concealed.

She had covered it over with silken sheen
  Away from the world's cold stare;
Her robes were fit for a queenly garb,
  She was loaded with diamonds rare.

She was wooed and won in her life's young spring,
  When the roses sweet were bringing
Fresh odors rare on the balmy air
  And the love birds sweetly singing;

But within the closet the ghastly form
  To the outward gaze concealed,
Though draped with velvet and costly gems,
  The skeleton stands revealed.

ON SENDING A BOUQUET OF AUTUMN ROSES.

TAKE a few more buds, dear Bertha;
  But I fear the end has come
For the lovely floral darlings
  That have beautified our home;
For last night the Frost King touched
    them,
  And his icy, chilling breath
Will despoil them of their
    beauty
  And consign them
    soon to death.

Soon their sweetness
    will have van-
    ished,
  Soon upon the garden bed
They will all, alas! be scattered
  And lie withered, cold, and dead;
But the spring will bring the roses,
  And in June some sunny morn
We may hail again their fragrance
  On the balmy breezes borne.

NEARING THE STREAM.

SOFT Twilight her curtains has drawn.
I sit in the gloaming,
And in the sweet gardens beyond
In fancy I'm roaming;
And fond, loving greetings I find
From dear ones departed,—
Their hearts beat as warm and as true
As when we first parted.

The dear, loving friends of my youth,
With love-beaming faces,
All meet me with tender caress
And dear, fond embraces.
From father and mother at home
I find joyful greeting,
And brother and sister there, too,—
How glad is the meeting!

That brother to manhood has grown
Who passed through death's portal
Ere I had embarked on life's sea
And entered the mortal;
Each heart knows its own, and the bliss
Of blest recognition

"They passed o'er the river of change
In life's fresh, sweet dawning."

Makes home seem more sweet and more dear
  In glorious fruition.

There are comrades I loved long ago
  In youth's dewy morning;
They passed o'er the river of change
  In life's fresh, sweet dawning;
They meet me all radiant with joy —
  No traces of sadness;
Sweet welcome I feel in each tone,
  My heart swells with gladness.

Each thorn that my footsteps have pressed
  Still brings the goal nearer;
The shadows that darken life's path
  Make heaven still dearer;
I 'm nearing the stream, and I see,
  With faith's open vision,
The dear ones of sweet long ago
  In gardens elysian.

BUILDING OUR HOME.

"'T is not the wide phylactery,
Nor stubborn taste nor stated prayers
That makes us saints; we judge the tree
By what it bears."

—*Alice Carey.*

Two souls came down to the river side,
And the angel Death, with his shadowy oar,
Safe carried them over the foaming tide
And bore them across to the further shore.

And as up to the pearly gates they passed,
One walked erect without sign of fear,
Which seemed to say, "I am safe at last,
And *sure* of gaining an entrance here."

The other with faltering footsteps trod,
And passed with a quiet, humble mien
As they traveled up to the courts of God,
Where an angel met them in silvery sheen.

"From whence did you come?" to the first he spoke.
"I came from the earth," he answered. bold;
"I have worn through life the Master's yoke,
And I'm coming to walk the streets of gold.

"I believed in Christ, and each Sabbath day
I was always present to hear the word;
I was always ready to speak or pray —
My voice in the temple was always heard."

"Professions and prayers," spoke the angel voice,
"Availeth naught; only righteous deeds,
Which have made the sorrowing heart rejoice,
Are counted in heaven — not empty creeds.

"Have you sought to succor the souls distressed?
Have you fed the hungry and clothed the poor?
Have lifted burdens from hearts oppressed
And given them aid from your bounteous store?"

He answered not, and the angel's face
Sought out a region all bleak and bare.
"Behold yon barren and rock-bound place!
The home you have builded awaits you there."

Then the other approached with downcast eyes,
And the angel said to him, "What did you
When you dwelt in the world below the skies?"
He answered: "I tried to be just and true.

"The Golden Rule was my beacon light,
And 'Love to man' was my only creed;
I defended what I deemed truth and right,
And the voice of conscience approved the deed;

"I strove to be upright, true, and kind,
And earnestly sought to do my best;
I aimed to be pure in heart and mind,
And I leave to the Infinite all the rest."

The pilgrim waited in patient mood,
While a beauteous landscape rose up to view,
Where a dwelling, embowered in roses, stood,
All fresh, as if bathed with the morning dew.

"The home you have built," said the angel then,
"By generous actions and noble deeds,
By charity, kindness, and love to men,
Has more availed you than prayers and creeds.

"You have given relief to the suffering poor.
Profession alone is an empty thing,
But love and mercy for aye endure,
And joy to each earth-born mortal bring."

Earlier Poems

Earlier Poems.

LINES TO A SCHOOLMATE.

I 'LL think of thee often when morn's golden light
Dispels all the darkness and shadows of night,
When the bright orb of day is ascending on high,
And the lark loudly warbles his notes in the sky.

When dewdrops are sparkling the flowerets among,
And garlands of beauty o'er nature are flung,
When loveliness lingers in woodland and glen,
And all things smile sweetly, I 'll think of thee then.

I 'll think of thee oft when in silence I rove,
And list to sweet music in forests and grove;
And often I 'll wish that you, too, could be there,
To join in my rambles, my solitude share.

I 'll think of the wood where together we 've strayed,
Our minds to refresh 'mid the forest's cool shade,
And there on the beauties of nature to gaze,
And list to the wild bird's sweet anthems of praise.

I 'll think of thee often when twilight's sweet hour,
With all the enchantment and charms of its power,
Steals soft o'er my spirit in music serene,
Inviting my thoughts to a world yet unseen.

I 'll think of thee oft when the stars sparkle bright,
And twilight gives way to the glory of night;
When fair Cynthia comes forth in her beauteous array,
And her soft, melting radiance drives darkness away.

Wherever I wander, by night or by day,
Thy image is with me and ever shall stay;
I 'll bear in my bosom sweet memories of thee;
I ask but one boon — 't is remembrance of me.

RETURN OF WINTER.

THE Frost King has breathed o'er the valley and plain;
The flow'rets have withered and died;
The forest is stripped of its foliage again
That stood in its beauty and pride;
The robin has ceased his sweet songs of delight;
The lark by his warble no more
Awakes us from sleep, and to labor invites,
As oft he has called us before.

The squirrel has ceased his wild, sportive glee;
His gambols are ended at last;
The bee seeks her hive, and, from labor set free,
Partakes of her winter repast;
The wind whistles by with an echo so drear,
The storm rages wildly again;
Like sad, mournful music it falls on the ear,
As if chanting some sorrowful strain.

No longer we ramble at twilight's soft hour,
As o'er us the stars faintly gleam,
And roam through the meadows, and pluck the wild flowers,
Or bathe in the cool, limpid stream;
Its murmurings are hushed 'mid the tempests that blow,
'T will soon with ice fetters be bound;

The earth will be wrapped in a mantle of snow,
And all things look cheerless around.

O Winter, stern Winter, I dread thy return,
I shrink from thy icy embrace;
From scenes bright and gay with reluctance I turn
To gaze on thy cold, cheerless face.
Give me the sweet spring, with its pure, fragrant breeze,
Its meadows, its groves, and its flowers,
The soft evening zephyrs that float through the trees,
And songs of the birds in the bowers.

"The earth will be wrapped in a mantle of snow."

LOVE.

Oh, what is love? 'T is silver beams
Of pure celestial light, which streams
From God's eternal throne;
A precious boon, to mortals given,
To guide our souls from earth to heaven,
To taste its joys unknown.

Oh, what is love? A golden chain,
Which man has tried to rend in vain;
A pure and priceless gem;
A cord which round our spirits twines,
A bright star which forever shines
In Heaven's own diadem.

'T is found alike in hut and hall;
The high, the low, the great, the small,
Have worshiped at its shrine;
The humble tent, the marble dome,
The poor man's cot, the rich man's home,
Have owned its power divine.

In every clime, in every land,
From Arctic's snow to Tropic's sand,
And sea-girt isles afar,

Affection's gems as radiant shine
As diamonds from a golden mine,
Bright as the morning star.

I ask not for complexion fair,
For ruby lips, or shining hair;
A generous, loving heart
Will live—while beauty will decay
And wealth takes wings and flies away—
And solid joy impart.

O love! thou pure, bright spirit, guide,
Whatever griefs or woes betide;
Shelter us 'neath thy wing;
Around us throw thy potent spell,
And let us drink, while here we dwell,
From thy perennial spring.

Be thou our guardian angel here,
While passing through this desert drear;
Rule thou our inmost soul;
And then in heaven, with sinless tongue,
Be love divine our ceaseless song,
While endless ages roll.

TO A LITTLE NEPHEW.

YOU 'RE now a little boy, Roy,
With spirit free from care,
And childhood's merry hours are fraught
With many pleasures rare;
But as the years go gliding by
You 'll pass to manhood's state,
And boys who grow up good and true
Will all as men be great.

True greatness don't consist, Roy,
In honors, wealth, or fame,
But he whose life is grand and good
In deeds as well as name,
Whose years are full of noble acts,
Though humble his estate,
Whose soul dishonor never stained,
That man is truly great.

Improve each shining hour, Roy,
In learning all you can
And laying stores of knowledge by;
For when you 've grown a man
You 'll need it all to guide your feet
In paths of truth and right,

That wisdom may with virtue blend
    And make your future bright.

To playmates all be kind, Roy,
    And always try to do
To every one just as you wish
    To have them do to you.
Let kindness dwell within your heart
    To every living thing,
For gentle deeds and accents mild
    Will sweet contentment bring.

TO A SCHOOLMATE—MISS T. E. J.

"If on this page some future year
Thine eye should chance to fall,
Then think of one who placed them here,"
And other days recall.

What pleasing recollections cling
'Round happy days of yore;
But each the sad reflection brings,
They shall be ours no more.

How often like a pleasant dream
Past visions of delight
Shed o'er our hearts a lingering gleam
Of pure, unfading light.

And, though we now must parted be,
Those scenes shall ne'er depart;
Forever shall thy memory
Be graven on my heart.

And now, one precious boon I crave
Of thee, my faithful friend:
Let me be in thy heart enshrined
Till mortal life shall end.

ANGEL GUARDIANS.

How sweet to feel, in hours of grief and sadness,
  When joy has fled and hope almost departs,
When hushed is every silvery note of gladness,
  The kindly sympathy of kindred hearts.

But oh, to know that pure and loving angels—
  What dearer, sweeter boon to mortals given?—
Can come to us as eloquent evangels
  Fresh from the verdant, flowery plains of heaven!

They come to guard, to comfort, and to cheer us,
  To guide our barque adown life's turbid stream;
When dangers threaten they are ever near us,
  Dispelling darkness by some radiant beam.

They come to tell us of the life immortal,
  The peaceful shore beyond this mortal breath,
To gild our pathway to the grave's dark portal,
  And bear us o'er the narrow stream of death.

This is not all of life—oh, sweet assurance!
  The blessed knowledge nerves our souls to bear
The varied ills of life with calm endurance,
  Till called by angels their delights to share,

To live and grow, expanding every power,—
  Fairer than spring shall every virtue beam,—
Fresh knowledge gain with every passing hour,
  And rise still higher in the life supreme.

Sweet summer-land, how oft I long to sever
  The chains that bind my fettered spirit here,
Thy beauty-blooming shores to roam forever,
  With happy dwellers in that higher sphere!